SUSTAINABILITY IN THE BUILT ENVIRONMENT

An introduction to its definition and measurement

Carol Atkinson, Alan Yates and Martin Wyatt

bre press

BRE is the UK's leading centre of expertise on the built environment, construction, energy use in buildings, fire prevention and control, and risk management. BRE Global is a part of the BRE Group, a world-leading research, consultancy, training, testing and certification organisation, delivering sustainability and innovation across the built environment and beyond. The BRE Group is wholly owned by the BRE Trust, a registered charity aiming to advance knowledge, innovation and communication in all matters concerning the built environment for the benefit of all. All BRE Group profits are passed to the BRE Trust to promote its charitable objectives.

BRE is committed to providing impartial and authoritative information on all aspects of the built environment. We make every effort to ensure the accuracy and quality of information and guidance when it is published. However, we can take no responsibility for the subsequent use of this information, nor for any errors or omissions it may contain.

BRE
Garston
Watford WD25 9XX
Tel: 01923 664000
Email: enquiries@bre.co.uk
www.bre.co.uk

BRE publications are available from
www.brebookshop.com
or
IHS BRE Press
Willoughby Road
Bracknell RG12 8FB
Tel: 01344 328038
Fax: 01344 328005
Email: brepress@ihs.com

Published by IHS BRE Press

Requests to copy any part of this publication should be made to the publisher:
IHS BRE Press
Garston
Watford WD25 9XX
Tel: 01923 664761
Email: brepress@ihs.com

Front cover photographs by Nick Clarke

Printed on paper sourced from responsibly managed forests

BR 502

First published 2009
ISBN 978-1-84806-084-5

ACKNOWLEDGEMENTS

The authors would like to thank the following BRE colleagues for their helpful comments and advice:
Jane Anderson
David Crowhurst
Emma Franklin
Jackie Innes
Christine Pout
Josephine Prior
Martin Townsend
Chris Watson

CONTENTS

1 INTRODUCTION

This report brings together current thinking on defining and measuring sustainability in the context of the built environment. It sets out concisely the key issues in this large and complex area. This subject is moving rapidly so it is not possible to be comprehensive, but this introductory 'primer' aims to provide a clear overview of the key issues and initiatives in the UK and internationally.

In this report, the Brundtland definition of sustainability is used: 'meeting the needs of the present without compromising the ability of future generations to meet their own needs' (Brundtland, 1987). Unfortunately, the simplicity of this definition belies what is a complex web of systems and cycles in science, economics, politics, ethics and engineering. Figure 1 illustrates the complex relationships between sustainability issues and the impacts of a building, and Table 1 lists some of the major environmental issues, many of which are inter-related.

Fortunately, pioneers of sustainability assessment in the built environment have devised practical ways of addressing sustainability measurement and delivery by focusing on the key issues in terms of:

- economic impacts,
- environmental impacts, and
- social impacts.

Through clarity, transparency, stakeholder engagement and peer review, the leading organisations are also attempting to achieve the objective of Brundtland. This requires that, as understanding improves, we identify and reconcile all of the key issues, which are inextricably interwoven. If we are to solve the problems of sustainability, we need 'numbers – not adjectives', and must base what we do on 'evidence not public relations' (MacKay, 2008).

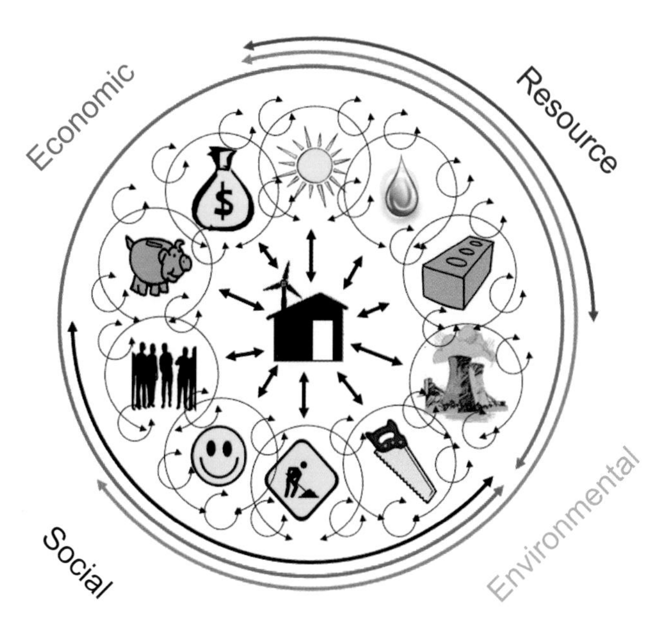

Figure 1: Relationships between sustainability issues are complex and interactive

In this publication, we do not attempt to tackle the many moral and ethical issues that arise from how we interact with the earth and the plants, animals and micro-organisms we share it with. For this reason, Table 1 focuses on the issues that have an impact on human beings.

***Table 1:* Key environmental issues**

Issue	Importance to humans
Climate change	Human activities have both direct and indirect impacts on the climate, which affects weather patterns and in turn has an impact on economic and social activity, as well as health, wellbeing, safety and demands for resource use.
Global warming	The earth's atmosphere acts as a greenhouse, trapping solar energy and heat. Without this process life could not exist on the planet. It is widely accepted that human activity, in particular the emission of CO_2, is acting to intensify this process, resulting in a gradual warming of the atmosphere. This is expected to result in significant climate change, although the level and nature of this change will vary considerably regionally. Global warming will cause considerable warming of the oceans, resulting in significant sea level rises, which could have major impacts on many centres of population and may alter the earth's albedo (ie the extent to which it reflects light from the sun).
Carbon emissions	Carbon emissions are a major cause of climate change, leading to major impacts on all aspects of human welfare.
Energy use	Energy is essential to human wellbeing, but unless its use is in balance with the capacity of the planet to absorb carbon emissions and waste heat, it will lead to major impacts on all aspects of human welfare. There are also related social and economic issues of energy security.
Global dimming	Global dimming is the reduction in irradiance at the earth's surface caused by absorption of radiation by particulates such as sulfate aerosols. It is thought to affect the world's hydrological cycles and create a cooling effect, which may have partially masked climate change. This has been linked to problems with water supply, crop failure and desertification.
Water resources	Water is essential to human wellbeing, but its use must be in balance with the local resources available to provide it and to process the resulting waste water.
Water availability	Human activities can reduce the availability of water through over-use and increased surface run-off, which can have significant impacts on human health and economic activity. There are also related social issues.
Over-use	Over-use can lead to: damage to ecosystems; loss of biodiversity; desertification; soil erosion; habitat destruction; risks to water security; and accelerated climate change due to energy use for purification, pumping and desalination processing.
Flooding	Flooding results from a loss of absorption capacity, which can directly result in damage to aquifers, property and economic activity. It can also result in contamination problems. Rapid run-off prevents replenishment of groundwater aquifers and hence leads to supply problems.
Salinisation	Over-extraction of groundwater for human use can lead to ingress of saline water into aquifers from the sea, which effectively prevents its continued use through increasing mineral levels. In addition, changes in rainfall can significantly alter the salinity of groundwater and seawater, impacting on both human health and biodiversity.
Water security	Both over-use and flooding can reduce water security. This has potentially serious political and health implications for many populations.
Pollution	Pollution can have a significant impact on human health, ecosystems and amenity value as a result of toxicity and acidity in the natural environment. It can also result in physical damage to property and infrastructure. The disposal of waste is a major cause of pollution to land, water and air and can have major impacts on human and ecological health.
Water pollution	Water is essential to human wellbeing and ecosystems, but run-off from industry and agriculture can damage both surface water and groundwater, in some cases making it toxic.
Marine pollution	Agricultural, industrial and human waste pollution of the sea has more complex interactions than simple freshwater pollution but, in addition to loss of fish stocks, it may be implicated in a loss of the ocean's capacity to absorb carbon – hence accelerating climate change.
Air pollution	Pollutants have been linked to rises in pulmonary disorders, and corrosive pollutants such as ozone and sulphurous oxides have damaged forests, fisheries and buildings. Some are leading to global dimming.
Acid rain	Emissions of nitrous and sulphurous oxides from the burning of fossil fuels result in high levels of acidity, which are carried to ground level in precipitation. This 'acid rain' can cause severe damage to the built environment and ecosystems, as well as having major impacts on the productivity of agriculture and forestry.
Soil pollution	Various industrial and agricultural byproducts can damage ecosystems and the food-generation capacity of soil, in some cases making it toxic.

Table 1: **Key environmental issues (continued)**

Issue	Importance to humans
Soil erosion/ flooding	Over-development of land can lead to heavy surface water run-off, causing flooding, pollution and soil erosion. These can be mitigated by appropriate construction and provision of features designed to reduce flows and retain water (sustainable drainage) or to retain soil through physical barriers, hedges and other planting.
Land take	Development often diverts land from other economically important activities or from leisure use. Waste also results in significant land take.
Land remediation	Reuse of land potentially brings it back into productive use and reduces the demands on other land resources.
Biodiversity	Many of our technological advances, especially in the fields of agriculture, medicine and alternative energy, depend on natural resources to provide a starting point. The loss of plant and animal species limits the future potential for medical and agricultural research. More intensive agriculture and development can have significant impacts on biodiversity.
Habitat destruction	Development can be very damaging to natural habitats. It impacts on the provision of leisure facilities and can lead to a loss of local biodiversity. Development can be carried out in a manner that limits damage or enhances ecological value.
Damage to ecosystems	Ecosystems are dynamic relationships of plants, animals, micro-organisms and the non-living environment. They can play many roles including pollination, protection against soil erosion, air quality improvement, climate regulation, water purification and regulation, waste treatment, and biological and disease control. While they often have some functional redundancy, serious damage or destruction of ecosystems can have major impacts on human health, agriculture (crop yield), climate and infrastructure.
Resource depletion	Almost all resources that we currently depend on are finite or otherwise limited. Loss of scarce resources could lead to a loss of technological capability and have harmful results on industry, economics and society in the future.
Ozone depletion	Although various protocols have agreed that chlorofluorocarbons (CFCs) and hydrochlorofluorocarbons (HCFCs) should no longer be produced, black market trading continues to be a problem, especially with existing systems, and is delaying the 'healing' of the ozone layer with all its adverse implications for human health and the natural environment.
Desertification	Reduction of land available to support human life leads to hunger, migration and pressure on economies.
Population	Population density has a major effect on demands for resources, including water and energy, and ecosystems, land use and pollution levels.

2 SUSTAINABILITY OF CONSTRUCTION

Construction and its products not only underpin all of the UK's economic activity, but also directly contribute some £100 billion (~10%) to the UK's gross domestic product (GDP) (Department for Business, Enterprise & Regulatory Reform, 2008; Office of National Statistics, 2008). By comparison, manufacturing contributes 16% of GDP, defence 4% and agriculture just 1%. UK construction can be 'world class' with modern buildings that are more comfortable and efficient, use more natural lighting, are better ventilated, more flexible, cheaper to heat (and cool) and use significantly less energy than existing buildings. In addition, the characteristics and performance of most basic construction materials are now well understood and long lifetimes can be achieved simultaneously with low maintenance costs.

The construction industry's success is further demonstrated by its substantial export earnings, some £10 billion per annum, arising particularly from the activities of constructors, engineers and architects who deliver high-quality buildings and infrastructure projects worldwide. Its design skills alone generate around £3.8 billion export income per annum through such high-profile projects as Madrid's Barajas Airport, Clarke Quay in Singapore, the Marbach Deutsches Literaturarchiv in Germany, Stonecutters Bridge in Hong Kong, Beijing National Stadium and many others.

This standing and economic strength have been achieved through the education and training of first-class design and construction professionals, underpinned by a substantial manufacturing base and a skilled workforce. The leading technological edge and continuous improvement in construction have been maintained primarily:

- by innovative manufacturers and constructors, and
- through codes and standards underpinned by many years of applied research and innovation provided by a network of applied research, technology and innovation organisations working collaboratively with government, industry and the universities.

Unfortunately, the high economic contribution of the construction industry and the built environment comes at high environmental cost with, for example, buildings accounting for around 45% of total UK greenhouse gas emissions (Pout and MacKenzie, 2005) and production of materials accounting for a further 10% (Office of National Statistics, 2008). While what constitutes sustainability is a complex issue requiring solutions from the fields of science, engineering, politics, economics, society and ethics, the UK has played a leading role in setting the pace for sustainable construction since 1990, when the BRE Environmental Assessment Method (BREEAM), the world's first environmental assessment method for buildings, was launched. BREEAM is the collaborative result of many years' development of codes, standards and toolkits by a network of organisations working with government, industry and the universities. In addition to the growing BREEAM family of standards, significant standards and tools include:

- CEEQUAL – the Civil Engineering Environmental Quality Assessment and Award Scheme,
- DQI – the Construction Industry Council's Design Quality Indicator,

- FSC – the Forest Stewardship Council, and
- WWF's One Planet Future.

Company tools such as Arup's SPeAR® have also done much to advance knowledge.

These codes and standards are designed to work with some or all parts of the UK's fragmented construction industry and at different stages in a construction product's life cycle. An overview of these and other publicly available tools is given in Section 3, while some of the best known international codes and standards are described in Section 4.

Regulatory and voluntary mechanisms should work together to encourage optimal performance. Regulation is a powerful and necessary tool for achieving sustainability targets, but it can only set a common base. Over time this can be raised to achieve major changes, but this must be at a pace that the majority of the industry can deliver. Voluntary codes and standards provide a means of encouraging industry leaders and innovators to go further and faster. Table 2 maps UK regulations and voluntary codes against key stages and activities in the construction life cycle. An example of the steady increase in minimum standards is the department of Communities and Local Government's (CLG) proposed 'road map to zero carbon domestic buildings' (Turner, 2008). Regulatory standards vary greatly across the world. This is the case even with highly developed economies such as those of Western Europe (including the UK) and North America. These variations in the statutory baseline arise from physical differences such as climate and construction technologies as well as cultural and political differences.

Construction and the management of the built environment involve many stakeholders. Relationships are complex and the tools, standards and guidance are usually focused on parts of this network. Figure 2 illustrates some of the major relationships against which they can be mapped. In reality, all stakeholders are impacted to some degree by most of the initiatives covered in this report, but this is often indirect. Any driver for change needs to target specific stakeholders while considering the impacts on others less directly affected. A 'one size fits all' tool is unlikely to meet any individual stakeholder's needs robustly.

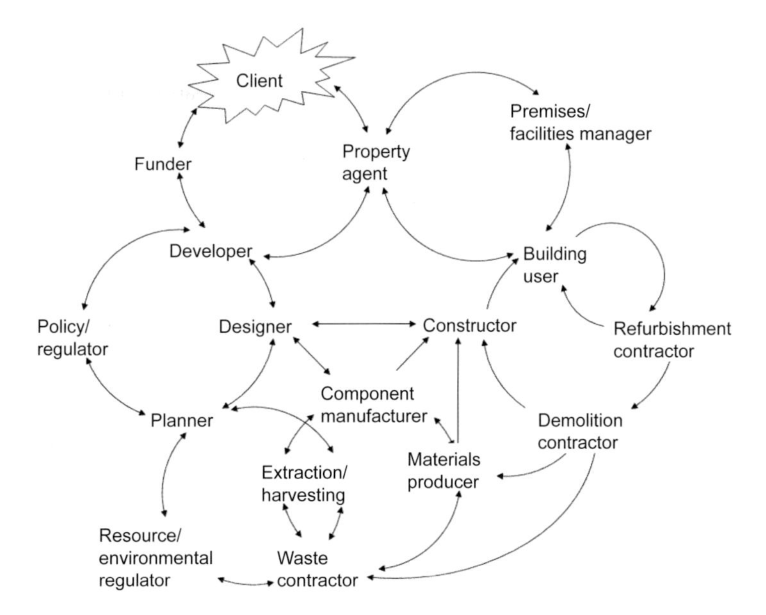

Figure 2: Relationships between parties in construction projects can be complex

Table 2: **Regulatory and voluntary framework for sustainable construction in the UK**

Stage	Principal sustainability regulations	Voluntary codes and standards
Planning and infrastructure	• Climate Change Act 2008 • Environmental Impact Assessment Directive 85/337/EEC (as amended by 97/11/EC) • Local Development Frameworks and Development Plans (numerous) • Planning Act 2008 • Planning and Compulsory Purchase Act 2004 (Planning Policy Statements (PPSs) (various) and Planning Policy Guidance Notes (PPGs) (various)) • Planning and Energy Act 2008 • Regional Spatial Strategies (various) • Strategic Environmental Assessment Directive 2001/42/EC	• BREEAM Buildings • BREEAM for Communities • CEEQUAL • Code for Sustainable Homes (CSH)
Design	• Building Regulations (all parts) • Disability Discrimination Act 1995 • Energy Performance of Buildings Directive 2002/91/EC – Energy Performance Certificates (EPCs) • Household Waste Recycling Act 2003	• BREEAM Buildings including BREEAM Energy Star • CEEQUAL • Code for Sustainable Homes (CSH) • Design Quality Indicators (DQIs) • Numerous specialist codes and standards • Numerous specialist regulations relating to building functions • Secured by Design • Standards developed under European Commission Mandate M350 (under development)
Materials and procurement	• Construction Products Directive • Convention on International Trade in Endangered Species of Wild Flora and Fauna (CITES) • European Emissions Trading Scheme 2005 • Marine Minerals Guidance Notes and Circulars • Minerals Planning Statements and Policy Guidance Notes • Registration, Evaluation, Authorisation and restriction of CHemicals (REACH) • Restriction of Hazardous Substances Directive (ROHS) • Waste Electrical and Electronic Equipment (WEEE) Directive and Regulations	• Environmental Profiles (LCA) • EU Eco-labelling Regulation (EC) 1980/2000 • ISO 14000/EMAS • Standards developed under European Commission Mandate M350 (under development) • Sustainable/responsible sourcing (FSC; BES 6001)
Construction	• Building Regulations (all parts) • Clean Air Act 1993 • Construction, Design and Management (CDM) Regulations 2007 • Control of Noise (Codes of Practice for Construction and Open Sites) (England) Order 2002 • Control of Pollution Act 1974 • Energy Performance of Buildings Directive 2002/91/EC – Energy Performance Certificates (EPCs) • Environment Act 1995 • Environmental Protection Act 1991 • Hazardous Waste Regulations 2005 and List of Wastes Regulations 2005 (both vary for England, Wales, Scotland and Northern Ireland) • Pollution Prevention and Control Act 1999 • Site Waste Management Plans Regulations 2008 • Waste Electrical and Electronic Equipment (WEEE) Directive and Regulations • Water fittings regulations (various) • Wildlife and Countryside Act 1981, (Amendment) Act 1991 and (Amendment) Regulations 2004	• BREEAM including BREEAM Energy Star; CEEQUAL; standards developed under European Commission Mandate M350 (under development) • Code for Sustainable Homes (CSH) • Considerate Constructors Scheme (and similar) • ISO 14000/EMAS • SMARTWaste

Table 2: **Regulatory and voluntary framework for sustainable construction in the UK (continued)**

Stage	Principal sustainability regulations	Voluntary codes and standards
Use	• Control of Pollution Act 1974 • Energy Performance of Buildings Directive 2002/91/EC – Display Energy Certificates (DECs) • Health and Safety at Work Act 1974 • Management of Health and Safety at Work Regulations 1999 • Regulatory Reform (Fire Safety) Order 2005 • Waste Regulations (various) • Water Industry Act 1991 • Workplace (Health, Safety and Welfare) Regulations 1992	• BREEAM in Use – Asset Rating, Operational Rating and Organisational Rating • Standards developed under European Commission Mandate M350 (under development)
Maintenance and refurbishment	• Energy Performance of Buildings Directive 2002/91/EC – Energy Performance Certificates (EPCs) • Workplace (Health, Safety and Welfare) Regulations 1992	• BREEAM for refurbishment (under development); standards developed under European Commission Mandate M350 (under development) • BREEAM in Use – Asset Rating
Property sale/ transfer	• Energy Performance of Buildings Directive 2002/91/EC – Energy Performance Certificates (EPCs) • Home/Building Information Pack Regulations (HIP/BIP)	
Demolition	• Clean Air Act 1993 • Construction, Design and Management (CDM) Regulations 2007 • Control of Noise (Codes of Practice for Construction and Open Sites) (England) Order 2002 • Environmental Protection Act 1991 • Hazardous Waste Regulations 2005 and List of Wastes Regulations 2005 (both vary for England, Wales, Scotland and Northern Ireland) • Health and Safety at Work Act 1974 • Management of Heath and Safety at Work Regulations 1999 • Pollution Prevention and Control Act 1999 • Waste Electrical and Electronic Equipment (WEEE) Directive and Regulations • Wildlife and Countryside Act 1981, (Amendment) Act 1991 and (Amendment) Regulations 2004. • Workplace (Health, Safety and Welfare) Regulations 1992	• ISO 14000/EMAS • Standards developed under European Commission Mandate M350 (under development)

Notes

1. This list of regulations and standards is included for illustrative purposes and is not exhaustive. Readers should check that they refer to current regulations in force in their region.
2. Further details of the above regulations can be found in Table 3.

Table 3: Websites for UK regulations and voluntary codes

Legislation, regulation, voluntary codes and standards	Website address
All enacted UK legislation (available from the Office of Public Sector Information; part of the National Archives)	*www.opsi.gov.uk*
BREEAM family of assessment methods	*www.breeam.org*
CEEQUAL	*www.ceequal.com*
Code for Sustainable Homes (CSH)	*www.communities.gov.uk*
Considerate Constructors Scheme	*www.considerateconstructorsscheme.org.uk*
Construction, Design and Management (CDM) Regulations 2007	*www.hse.gov.uk*
Construction Products Directive	*www.communities.gov.uk*
Control of Noise (Codes of Practice for Construction and Open Sites) (England) Order 2002	*www.opsi.gov.uk*
Convention on International Trade in Endangered Species of Wild Fauna and Flora (CITES)	*www.cites.org*
Design Quality Indicators (DQIs)	*www.dqi.org.uk*
Disability Discrimination Act 1995	*www.dwp.gov.uk*
Energy Performance of Buildings Directive 2002/91/EC – Energy Performance Certificates (EPCs)	*www.communities.gov.uk*
Environmental Profiles (Life Cycle Assessment)	*www.greenbooklive.com*
EU Eco-labelling Regulation (EC) 1980/2000	*http://ec.europa.eu/environment/ecolabel/documents/pm_regulation_en.htm*
European Directives (all published in the Official Journal of the European Communities)	*http://eur-lex.europa.eu/JOIndex.do*
European Emissions Trading Scheme 2005	*http://ec.europa.eu/environment/climat/emission.htm*
Framework standard for the Responsible Sourcing of Construction Products, BES 6001 (available from BRE Global Ltd)	*www.bre.co.uk*
Hazardous Waste Regulations (England and Wales) 2005 (other regulations apply in Scotland and Northern Ireland)	*www.opsi.gov.uk*
Home/Building Information Pack Regulations (HIP/BIP)	*www.communities.gov.uk*
Household Waste Recycling Act 2003	*www.defra.gov.uk*
International standards published by the British Standards Institution (BSI), including ISO 9000 (BS EN ISO 9001:2004) and ISO 14000 (BS EN ISO 14001:2004)	*www.bsi-global.com*
List of Wastes Regulations (England) 2005 (other regulations apply in Wales, Scotland and Northern Ireland)	*www.opsi.gov.uk*
Management of Health and Safety at Work Regulations 1999	*www.opsi.gov.uk*
Registration, Evaluation, Authorisation and restriction of CHemicals (REACH)	*www.hse.gov.uk/reach*
Regulatory Reform (Fire Safety) Order 2005	*www.opsi.gov.uk*
Restriction of Hazardous Substances Directive (ROHS)	*www.berr.gov.uk*
Secured by Design	*www.securedbydesign.com*
SMARTWaste	*www.smartwaste.co.uk*
Standards developed under European Commission Mandate M350 (under development) – Integrated Environmental Performance of Buildings (European platform on Life Cycle Assessment)	*http://lca.jrc.ec.europa.eu*
UK government Planning Acts, Building Acts and all their supporting regulations, policy statements and policy guidance	*www.planningportal.gov.uk*
Waste Electrical and Electronic Equipment (WEEE) Directive and Regulations	*www.berr.gov.uk*
Waste regulations pertaining to the built environment in use (various)	*www.opsi.gov.uk*
Water fittings regulations	*www.defra.gov.uk*
Workplace (Health, Safety and Welfare) Regulations 1992	*www.opsi.gov.uk*

3 UK TOOLS FOR MEASURING AND RATING SUSTAINABILITY OF BUILDING MATERIALS AND PRODUCTS

3.1 BREEAM – BRE Environmental Assessment Method

www.breeam.org

Winner of the award for 'the worldwide best program for environmental assessment' at the World Sustainable Building Conference (2005) in Tokyo, BREEAM was the first and continues to be the world's leading environmental rating and assessment method for buildings.

Launched in 1990, BREEAM is based on many years of research, scientific and market analysis. It assesses the environmental impacts of buildings in terms of: energy; transport; health and wellbeing; water; materials; waste; pollution; land use and site ecology; and management (BRE, 1990–2008). The method has been regularly enhanced since this date to ensure that it both reflects current regulations, standards and industry practices while providing an incentive to achieve the most out of a development project. There are a number of related and integrated tools in the BREEAM suite. Figure 3 is a schematic diagram of the BREEAM 'family' of standards and tools, which now cover almost all stages in a building's life cycle.

This now includes operation and management, through BREEAM in Use, and broader infrastructure and planning issues, through BREEAM Communities. The assessment method covers all buildings through sector-specific versions relating to building types, public sector procurers or a bespoke assessment process, which provides a filtering of criteria to ensure relevance to the project in hand while maintaining a common overall standard.

Since the introduction of BREEAM in 1990, the standard has been kept ahead of, but in step with, UK sustainability regulations and ramped up as fast as the market will bear (BRE, 1990–2008). As a voluntary standard, BREEAM must keep in step with the economic costs of higher sustainability and property market expectations. Working closely with the UK government, BREEAM has also helped the government test the market impacts of its policy imperatives.

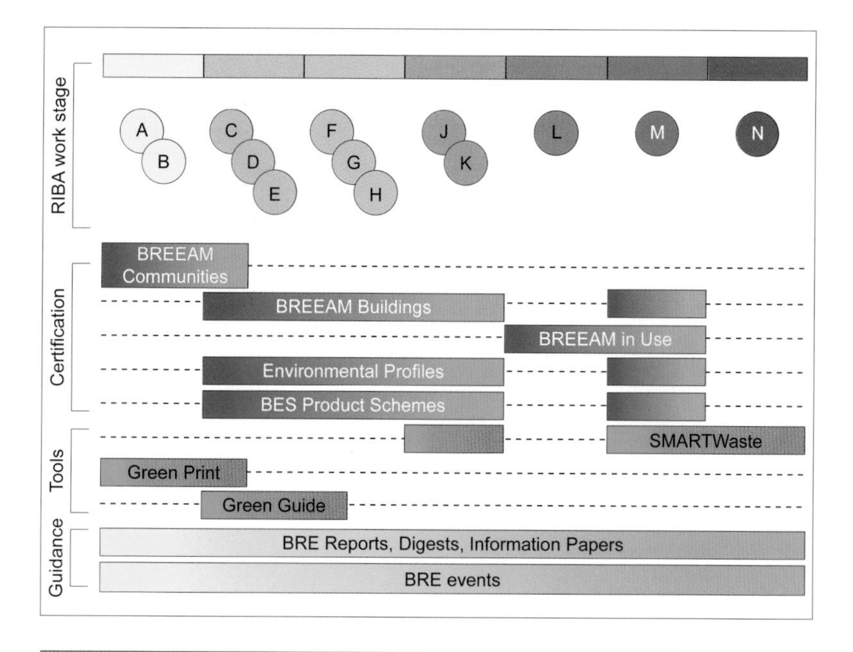

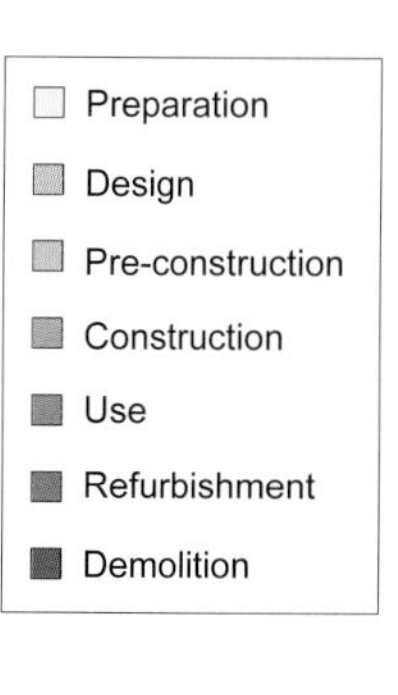

Figure 3: The BREEAM family of standards and tools

For example, the first domestic BREEAM in 1991 gave credits for smoke detectors, which became mandatory in the UK in 1997; recently, in autumn 2008, BREEAM Energy Star was introduced, which will be kept aligned with the department of Communities and Local Government's (CLG) proposed 'road map to zero carbon domestic buildings' (Turner, 2008), to encourage developers to meet the new building targets for zero carbon before 2016 (domestic) and 2019 (non-domestic) (CLG, 2007; Milne, 2008). The only exception is where experts consider policy to be 'ill-advised' – ie it will not work – in which case they have always tried to warn officials and help them alert their ministers.

Post-construction assessments are now mandatory for final certification. To encourage designers and builders to innovate and enable lessons learned to be promulgated, an 'Outstanding' rating for exemplar buildings has also been introduced with 'Innovation' credits to recognise beneficial sustainability aspects of the design that are not covered under the standard criteria.

Technical experts from BRE, academia and industry stakeholder groups develop the standards, but suggestions for improvement are welcomed from anyone. An informal users' group, with regular newsletters and updates, is being opened up more widely to enable easier uptake of higher sustainability standards and to facilitate feedback. All BREEAM assessments are carried out under licence by non-BRE companies and personnel – some 10,000 non-BRE personnel are thought to earn some or all of their living through BREEAM in the UK. Figure 4 sets out the operating structures and governance arrangements for the BREEAM family.

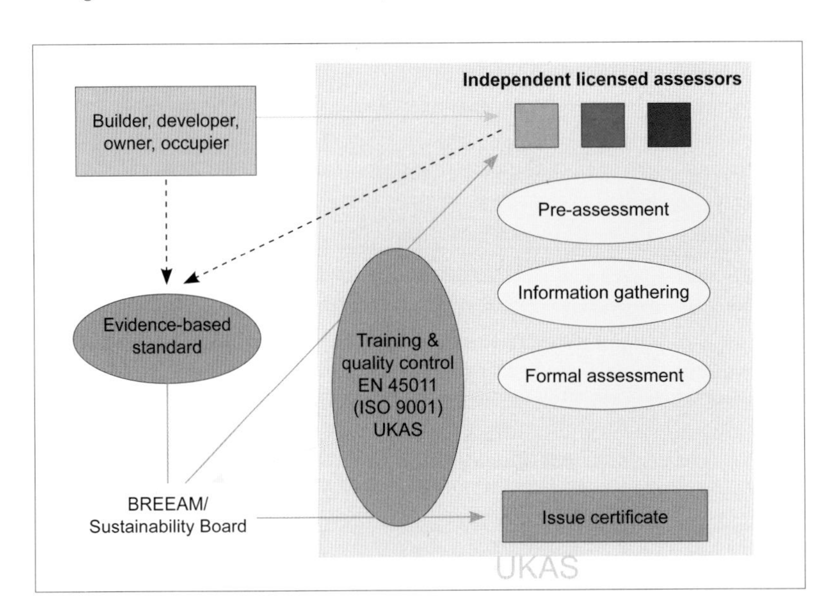

Figure 4: Operating roles and relationships within BREEAM

BREEAM has been recognised, even in popular publications, as a powerful force for greener building (Henson, 2006). The method is estimated to have resulted in savings of approximately 4.5 million tonnes of CO_2 over statutory requirements since its inception in 1990. This equates to the annual emissions from 40,000 average UK homes over basic Building Regulations levels or the total emissions from over 750,000 homes since 1990 (unpublished calculation based on 2005 BREEAM Offices assessment data).

While BREEAM is owned by the research and education charity the BRE Trust, the content and operation of BREEAM is overseen by a Sustainability Board dominated by independent stakeholders who neither pay nor are paid for the privilege and responsibility. The development and operation of BREEAM along with the training, operation and quality management of the assessor network are all accredited by the United Kingdom Accreditation Service (UKAS) to ensure independence, impartiality, probity and robustness.

3.2 CSH – Code for Sustainable Homes

www.planningportal.gov.uk

CLG launched the CSH in December 2006 with the method being released in April 2007. This method is based on BRE's EcoHomes version of the BREEAM methodology adapted to relate closely to Building Regulations and government policy. The method is owned by CLG and operated on its behalf by BRE Global, which also acts as technical advisor. BRE Global is contracted to operate the scheme and also license other operators to offer certification under the scheme.

The method sets mandatory minimum standards against energy, water, construction and household waste, materials and lifetime homes that relate to key government targets and policies. It has six potential star ratings.

Since October 2007, Level 6 has required a net zero carbon solution achieved through a private wire arrangement to bring the CSH in line with the UK Treasury's definition of a zero carbon dwelling used to determine eligibility for the exemption of stamp duty land tax. This definition has resulted in concerns over its practicality as it precludes any use of community or off-site-based energy systems. BRE Global, the UK Green Building Council (UKGBC) and others have recommended that this definition be expanded to encompass the use of such systems once cost-effective on-site options have been exhausted where they can be considered additional to existing commitments to renewable energy provision. CLG has yet to make a decision on any changes.

3.3 DQI – Design Quality Indicator

www.dqi.org.uk

www.dqi.org.uk/schools

The DQI is a process for evaluating the design quality of buildings; it can be used by everyone involved in the development process to contribute to improving the quality of our built environment. DQIs provide a generic toolkit that can be used with all types of building. There is also a version specifically aimed at school buildings. DQIs provide a framework for understanding quality priorities, setting targets and monitoring performance against them to evaluate design quality. They do not set specific performance levels, but provide an effective self-assessment process for use within the design process. Figure 5 shows the three elements of DQI – build quality, functionality and impact.

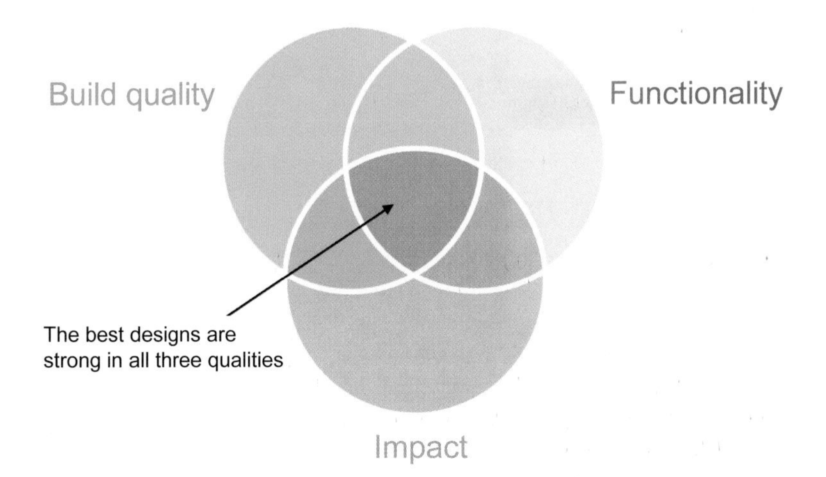

Figure 5: The three elements of DQI

The DQI process revolves around a workshop where the facilitator and design team members develop a set of project-specific targets against each of the DQIs. These targets are used to inform the briefing, design or management process. A post-construction review process is incorporated to allow follow-through of targets and feedback post-project. Training is provided for facilitators.

The development of DQIs was led by the Construction Industry Council (CIC) with input from a number of design consultancies and sponsors including the Commission for Architecture and the Built Environment (CABE), Office of Government Commerce (OGC) and Association of Building Engineers (ABE). The method is owned by CIC and is currently being adapted under licence in North America by DQI USA.

www.ceequal.com

3.4 CEEQUAL – Civil Engineering Environmental Quality Assessment and Award Scheme

CEEQUAL provides a generic assessment of the environmental quality of the design and construction of major civil engineering projects, and as such complements BREEAM, which focuses on buildings and communities. Developed by a number of major industry partners under the auspices of the Institution of Civil Engineers (ICE), it is based on the structure of BREEAM and was launched in 2004.

The method promotes consideration of sustainability issues throughout the procurement process and covers the following issues:

- Project management
- Land use
- Landscape
- Ecology and biodiversity
- The historic environment
- Water
- Energy and carbon
- Use of materials
- Waste
- Transport
- Effects on neighbours
- Relations with the local community and other stakeholders

CEEQUAL focuses on the actions undertaken to ensure that environmental quality is built into the design and construction processes. Unlike BREEAM, it does not reward or benchmark against specific measured performance levels as these vary between project types.

Current versions of the method are designed for use on projects with clearly defined boundaries. A term contracts version of the scheme is being developed that will be aimed at the assessment and recognition of environmental performance on maintenance or minor works over a period of time.

As with BREEAM, CEEQUAL builds on the current regulatory framework and provides guidance and environmental good practice in civil engineering projects. It provides a protocol for assessing, benchmarking and 'labelling' the sustainability performance of such projects.

Within CEEQUAL, six awards are available to recognise the roles of different stakeholders and stages in the procurement of a project:

- Whole Project Award (WPA), applied for jointly by or on behalf of the client, designer and principal contractor(s)
- WPA with an Interim Client & Design Award, where the stage in the design process at which the interim assessment is undertaken can be chosen by the applicant to best suit their needs and procurement process
- Client & Design Award, applied for jointly by the client and designer before construction has started
- Design Award, applied for by the principal designer
- Construction Award, applied for by the principal contractor(s)
- Design & Build Award, for project teams that do not include the client on design-and-construct and other partnership contracts

Assessments are carried out internally within a range of procurement design or construction organisations. To ensure independence, these assessments are independently audited by verifiers trained and licensed by CEEQUAL Ltd. The method includes a range of credit areas that lie outside the scope of specific projects. For this reason, the assessment process includes a scoping stage where these credits can be removed if they are not relevant. The assessment process is as follows:

1. Scoping – assessor and verifier
2. Assessment – assessor
3. Submission – assessor
4. Verification – verifier
5. Certification – CEEQUAL Ltd

The scheme is owned and operated through CEEQUAL Ltd, to which the Construction Industry Research and Information Association (CIRIA) and Crane Environmental are contracted to administer the company and scheme. CEEQUAL Ltd is supported by, amongst others, ICE, the Civil Engineering Contractors Association and the Association for Consultancy and Engineering. CEEQUAL is not covered by any form of external accreditation.

3.5 Other tools

3.5.1 Environmental Profiles

Demand for advice and information on the environmental impacts of products in the early 1990s led to an EU eco-labelling scheme, which was originally intended to apply to both consumer and construction products. Work by BRE demonstrated that the EU regulation was unsuitable for building products where it is essential (but very difficult) to take the application and whole-life performance into account (Atkinson and Butlin, 1993). This led to further work by BRE with the Construction Products Association (CPA) to develop an environmental profiling methodology suitable for building products. This was initially based on the SETAC principles of Life Cycle Assessment (LCA) (see Section 4.11), but subsequently BRE Global has been working to incorporate and ensure compatibility with the emerging ISO standards and European Commission Mandate M350 (see Section 4.2).

LCAs of manufacturing processes apportion energy, water, waste and raw material costs and impacts to the production of a range of products and co-products. While this sounds quite simple, there are a lot of contentious issues relating to how these inputs and outputs are calculated and allocated to co-products, how waste is defined and how future recycling should be accounted for. For example, the production of iron for the steel industry also produces slag, some of which can be used as fill, aggregate and in blended cements. If environmental impact allocations are based on the mass of the resulting products, most of the impacts become associated with the aggregate (Atkinson and Butlin, 1993). For this reason, value allocation has become more common, but this in turn becomes very complicated because of rapid changes in prices in the commodity market. In the absence of a clear basis for calculation, it is vital that there is transparency in the rules and education of specifiers in the use of Environmental Profiles and LCA data.

As explained above, LCA is complex, and the science and economics are still being developed and debated. To make the results of the work accessible, BRE Global has collaborated with CPA and others to produce Environmental Profiles in a standardised form using standard assumptions and algorithms, which enable specifiers to identify the major environmental impacts of a product at the point where the product leaves the factory gate. Specialist practitioners can use the data and different calculation algorithms to assess impacts under different conditions and different allocations of environmental weighting.

3.5.2 The Green Guide to Specification

www.thegreenguide.org.uk

With the help of CPA and others, BRE has produced *The Green Guide to Specification*, which provides simple environmental ratings of construction elements based on LCAs (Anderson *et al*, 2002). The fourth edition of

The Green Guide to Specification is due for publication in spring 2009, and an online version is also available. It is designed for use by designers and specifiers who:

- want to minimise the environmental impacts of buildings, and/or
- need to provide evidence for BREEAM assessments of buildings they are designing or procuring.

http://envestv2.bre.co.uk

3.5.3 ENVEST

ENVEST is a specialist tool that enables specifiers to aggregate profile and whole-life data to better assess the environmental impacts of material and system specifications.

3.6 RSM – Responsible Sourcing of construction Materials

BRE – with CPA, the British Standards Institution (BSI) and others – has developed and published a framework standard for the Responsible Sourcing of Construction Products, BES 6001 (BRE Global, 2008). Key players in the construction materials and components sector are currently working with BRE to develop BES 6001-compliant, sector-specific standards. See Section 4.12 for more details on RSM standards.

4 THE INTERNATIONAL STANDARD SCENE

4.1 ISO 14001 – Environmental Management Systems

International Standard ISO 14001 (BS EN ISO 14001:2004; British Standards Institution, 2004) specifies requirements for an Environmental Management System to enable an organisation to develop and implement a policy and objectives that take into account legal and other requirements to which the organisation subscribes, and information about significant environmental aspects.

Because the standard is about how a company manages and improves its processes, ISO 14001 does not enable benchmarking and comparison between different buildings, processes and organisations. However, it does provide a useful framework for managers designing their systems.

4.2 European Commission Mandate M350 – Integrated Environmental Performance of Buildings

The European Commission has mandated the European Committee for Standardisation (CEN) to develop a suite of standards (Mandate M350) for the 'integrated assessment of environmental performance of buildings' based on a life cycle approach (European Commission, 2004). The standards are intended to provide a voluntary method for delivery of environmental information that supports the construction of sustainable works, including new and existing buildings (not all construction works will be included). Specific areas covered include frameworks for the assessment of:

- environmental performance (prEN 15643-2),
- social performance (prEN 15643-3),
- economic performance (prEN 15643-4), and
- a general framework integration of these (prEN 15643-1).

These operate alongside rules for calculating and reporting:

- Environmental Product Declarations (EPDs) for construction materials (prEN 15804), which use a methodology similar to that used for BRE Environmental Profiles, and
- aggregated data from EPDs and other information – for example, energy performance data derived from National Calculation Methodologies such as SAP (BRE, 2005) and SBEM (BRE, 2006) – to produce a table of environmental impacts for a building across its whole life (similar in approach to the methodology used in BRE's ENVEST tool).

The standards will only describe methodologies for assessment; they specifically do not provide or attempt to prescribe benchmarks or levels of performance.

4.3 Green Globes

www.greenglobes.com

Green Globes is an environmental assessment and certification scheme based on BREEAM and includes self assessment with independent third-party verification. It is owned and operated by the Building Owners and Managers Association (BOMA) in Canada and the Green Building Initiative (GBI) in the USA.

The tool is available through subscription online. The service is designed primarily to provide interactive support and assessment to medium and small developments, although it is also widely used by property developers and managers on larger developments.

The method applies to both new build and existing buildings. Processes and development are accredited by the American National Standards Institute (ANSI).

www.breeam.org

4.4 BREEAM International

A comparison of environmental assessment methods found that none of them 'travelled' well without adaptation (Saunders, 2008) and that their use in different climates and regulatory regimes can undermine their validity.

To overcome such problems, BRE has developed an international version of BREEAM, BREEAM International, which will be open source and can be tailored to local conditions including climate, regulations and markets. In this way, BREEAM International should provide an accessible framework for assessment of sustainability and help to ensure a degree of comparability and 'stretch' based on local conditions.

The method sets out a core methodology and establishes the requirements for tailoring this to the local context. It includes default factors for environmental weightings where it is not possible or cost effective to develop local ones. Tailoring of the methodology can be done locally with or without BRE Global's help, but users of BREEAM International will be encouraged to share learning and knowledge and feed it back into the core standard development process.

www.usgbc.org/leed

4.5 LEED – Leadership in Energy and Environmental Design

LEED was launched in 1998 and was adapted from the UK BREEAM method to meet the needs of the USA. It was developed by the US Green Building Council (USGBC) to improve the way that the US construction industry addressed sustainability by providing a simple, easy-to-use label.

There are six versions of LEED in operation and a further five versions in preparation. Each version has four ratings – Certified (26–32 points), Silver (33–38 points), Gold (39–51 points) and Platinum (52–69 points) – based on the total number of credits that are achieved together with baseline performance in key areas. In addition, a number of mandatory requirements must be achieved before a rating can be awarded. These are not scored in the method.

LEED currently has the following versions:

- LEED for New Commercial Construction and Major Renovations
- LEED for Existing Buildings: Operation and Maintenance
- LEED for Commercial Interiors
- LEED for Core and Shell Development
- LEED for Homes
- LEED for Schools: New Construction and Major Renovations
- LEED for Neighbourhood Development (in preparation)
- LEED for Retail: New Construction (in preparation)
- LEED for Healthcare (in preparation)
- LEED Application Guide for Laboratories (in preparation)

The current LEED for New Commercial Construction and Major Renovations (version 2.2) was launched in 2005. This version is used throughout the design and construction phase, but the actual label (certificate) is only available once construction is completed.

The project team compiles the documentation required for the assessment. A trained assessor is therefore not required, although there is a credit available for appointing a LEED Accredited Professional (LEED AP) as part of the design team. Once all the documentation has been compiled, it is submitted to USGBC, which reviews the evidence and calculates the score. The project team has the opportunity to dispute the final score prior to USGBC issuing a certificate and a plaque with the rating on it.

There are no explicit weightings included within LEED. Individual credits are all worth one point (and where there are multiple performance levels each level is worth one point). The value of each issue is therefore dependent on the number of steps in the assessment criteria relating to actions that the project team can take rather than any measure of a reduction in environmental impacts resulting from the actions taken.

Currently, LEED uses a checklist approach to assess the embodied impact of the materials. This is an over-simplification, which leads to potential inaccuracies. USGBC is developing an approach that will bring the assessment of materials more into line with the Ecopoints/Green Guide method developed by BRE (Anderson *et al*, 2002).

LEED is developed by USGBC through a committee structure with representatives drawn from USGBC's membership. This allows representation from a wide range of sectoral interest groups, but leaves the process at some risk of bias to specific groups without course to redress from those impacted.

Until recently, all certification was carried out by USGBC. To overcome growing capacity difficulties, USGBC has recently established its own independent certification body, which in turn accredits a small number of external bodies to carry out certification under the scheme alongside USGBC. Assessments are apportioned on a round-robin basis. USGBC is accredited by ANSI as a standards developer (which covers the technical development of the standards). This accreditation does not cover the operation of the scheme.

Technical requirements relate to US standards, which vary from those in many other countries. This makes a direct comparison of credits between schemes difficult.

4.6 Green Star

www.gbca.org.au/green-star/

The first version of Green Star was developed for Australia in 2003 in a partnership between Sinclair Knight Merz and BRE. As BREEAM was used as the basis of the Green Star methodology, the two methods are very similar. However, adaptations have been made to reflect the various differences between Australia and the UK, such as the climate, local environment and standard practices in the construction industry. The environmental weightings have been re-evaluated to the Australian context using a similar approach to that adopted in the UK. Since the initial launch of Green Star, the Green Building Council of Australia (GBCA) has also adapted the assessment methodology to make the delivery mechanism more akin to the LEED approach where data are collected by the design team with verification prior to certification by GBCA.

There are currently seven versions of the methodology:

- Green Star – Office Design v3
- Green Star – Office As Built v3
- Green Star – Office Design v2
- Green Star – Office As Built v2
- Green Star – Office Interiors v1.1
- Green Star – Retail Centre v1
- Green Star – Education v1

Green Star can be used by any member of a design team or wider project team to provide a self assessment. No resulting score can be publicised unless the Green Star assessment is certified by GBCA. A third-party assessment panel is used to validate the self-assessment rating and recommend, or oppose, a Green Star-certified rating. Certification will only be awarded if a project achieves a score of at least 45 (four stars).

The following Green Star-certified ratings are available:

- Four-star rating (score 45–59) signifies 'best practice'
- Five-star rating (score 60–74) signifies 'Australian excellence'
- Six-star rating (score 75–100) signifies 'world leadership'

Green Star is developed and operated under a similar governance structure to LEED.

www.ibec.or.jp/CASBEE/english

www.greenbuilding.ca

4.7 CASBEE – Comprehensive Assessment System for Building Environmental Efficiency

CASBEE was launched in 2004 by the Japan Sustainable Building Consortium. The methodology is used to calculate a Building Environmental Efficiency (BEE) score that distinguishes between environmental load reduction and building quality performance. This was adapted from the approach first developed by the International Initiative for a Sustainable Built Environment (iiSBE) in the form of GBTool.

There are four versions of CASBEE:

- CASBEE for Pre-Design
- CASBEE for New Construction
- CASBEE for Existing Buildings
- CASBEE for Renovation

Under CASBEE, all building permit applicants must submit the required data, part of which are displayed on a public website. CASBEE is marketed primarily as a 'self-assessment check system' to permit users to raise the environmental performance of buildings under consideration. It can also be used as a labelling system, if the assessment is verified by a third party.

CASBEE is a complex calculation methodology. It uses weightings to balance the value of addressing issues with the number of measures recognised in the method. The more measures available to improve environmental performance, the more credits can be developed, but this does not necessarily reflect the environmental impact of addressing the issues. However, the weightings applied to CASBEE are much more complex than BREEAM, LEED or Green Star.

Weightings are applied to each category, which include 'indoor environment', 'outdoor environment onsite', 'energy' and 'resources and materials'. In each category, there are headline issues such as 'serviceability', 'lighting and illumination' and 'building thermal load', to which another layer of weightings is applied. Under these headline issues, there are individual issues including 'noise', 'ventilation' and 'use of recycled materials', which are also weighted. A final layer of weightings is applied to the sub-issues grouped under each of the individual issues. The sub-issues include 'ventilation rate', 'CO_2 monitoring' and 'adaptability of floor plate'.

All the issues are split into two basic types: quality measures (Q) and load reduction measures (LR). Once the assessment has been carried out, the final BEE score is calculated.

www.worldgbc.org

4.8 World GBC – World Green Building Council

World GBC is a union of national councils whose mission is to accelerate the transformation of the global built environment towards sustainability. The current member nations of the council represent over 50% of global construction activity, and reach more than 15,000 companies and organisations worldwide.

World GBC members are leading the movement that is globalising environmentally and socially responsible building practices. It aims to rapidly build an international coalition that represents the entire global property industry.

World GBC provides leadership and a global forum to accelerate market transformation from traditional, inefficient building practices to new-generation, high-performance buildings. This is a critical response strategy for cities and countries worldwide to their national and international commitments to reduce carbon emissions and redress other environmental impacts.

World GBC is a business-led coalition. Green building councils are consensus-based, not-for-profit organisations with no private ownership and diverse and integrated representation from all sectors of the property industry. They see business as a powerful solution-provider, and are working to improve frameworks that harness business's ability to deliver.

4.9 UNEP – United Nations Environment Programme

www.unep.org

UNEP is the United Nation's (UN) designated entity for addressing environmental issues at the global and regional level. The UN confers powers upon a designated entity in relation to the administration of UN sanctions and enforcement laws. Its mandate is to coordinate the development of environmental policy consensus by keeping the global environment under review and bringing emerging issues to the attention of governments and the international community for action. The mandate and objectives of UNEP emanate from UN General Assembly Resolution 2997 (XXVII) of 15 December 1972; Agenda 21, adopted at the UN Conference on Environment and Development (the Earth Summit) in 1992; the Nairobi Declaration on the Role and Mandate of UNEP, adopted by the UNEP Governing Council in 1997; the Malmö Ministerial Declaration and the UN Millennium Declaration, adopted in 2000; and recommendations related to international environmental governance approved by the 2002 World Summit on Sustainable Development and the United Nations 2005 World Summit. The UNEP Governing Council reports to the UN General Assembly through the Economic and Social Council. Its 58 members are elected by the General Assembly for four-year terms. An overview of UNEP can be found at *www.unep.org/resources/gov/overview.asp*.

UNEP is currently running a sustainable Buildings and Construction Initiative. This group has the following remit:

1. To promote improved support mechanisms for energy efficiency in buildings under the Kyoto Protocol.
2. To identify and support the adoption of policy tools that use a life cycle approach to investment in the building sector.
3. To develop benchmarks for sustainable buildings.

BRE is a full member of the group, focusing on the development of benchmarks for sustainable buildings. The group is seeking to contribute to the ongoing development of globally acknowledged benchmarks and is seeking as much common ground as possible while recognising that standards must be applied at a national level rather than globally. The group is not seeking to replace or compete with existing systems that are seen as positive.

4.10 SBA – Sustainable Buildings Alliance

www.sustainablebuildingsalliance.org

In 2007, SBA was founded by a group of leading organisations with a shared commitment to sustainable initiatives. The group is expanding and is committed to raising awareness amongst owners and occupants of the practical choices open to them in the design, construction and operation of their buildings and sharing experiences in promoting this agenda.

4.11 SETAC – Society of Environmental Toxicology and Chemistry

www.setac.org

SETAC was the first organisation to publish a framework for LCA (SETAC and SETAC Foundation for Environmental Education, 1991). LCA identifies and calculates all the inputs (energy and raw materials) and outputs (products and waste) for a defined product or system, providing information that can be used to interpret the effects on the environment.

SETAC is a not-for-profit, worldwide, professional organisation comprising individuals and institutions dedicated to the study, analysis and solution of environmental problems, the management and regulation of natural resources, research and development, and environmental education.

SETAC's mission is to support the development of principles and practices for protection, enhancement and management of sustainable environmental quality and ecosystem integrity. SETAC fulfils this mission through the advancement and application of scientific research related to contaminants and other stressors in the environment, education in the environmental sciences and the use of science in environmental policy and decision-making.

www.pefc.co.uk

www.csa.ca

www.sfiprogram.org

4.12 Supply chain management

Many parts of the construction materials supply sector are under increasing pressure to demonstrate sound environmental management at key stages in the supply chain. This is resulting in the development of standards and schemes for the responsible sourcing of materials (see Section 3.6) (BRE Global, 2008). This is a rapidly changing field, which started in the timber supply sector, where the Forest Stewardship Council (FSC) scheme was the first to provide a degree of robustness and rigour. Other schemes have followed suit in this sector, as a result of both marketplace and policy pressures, although their scope is not the same as that of FSC. These include the Programme for the Endorsement of Forest Certification (PEFC), Canadian Standards Association's (CSA) National Standard for Sustainable Forest Management and the Sustainable Forestry Initiative (SFI®).

www.fsc.org

4.12.1 FSC – Forest Stewardship Council

FSC is an independent, non-government, not-for-profit organisation established to promote the responsible management of the world's forests through certification against FSC standards. It is an international association whose members represent environmental and social groups, the timber trade and the forestry profession, indigenous people's organisations, responsible corporations, community forestry groups and forest product certification organisations from around the world. Its governance is built on principles of participation, democracy and equity, and its standards and policies are built on the following 10 principles*:

1. Compliance with laws and FSC principles – to cover all national and international laws and treaties/agreements to which the country is a signatory.
2. Tenure and use rights and responsibilities – to be fully established and documented.
3. Indigenous people's rights – to recognise and respect indigenous rights to land and resources.
4. Community relations and workers' rights – to maintain, enhance and respect long-term relationships with communities and workers.
5. Benefits from the forest – to promote efficient use of forest resources to ensure economic viability and a wide range of environmental and social benefits.
6. Environmental impact – to conserve biodiversity and forest resources.
7. Management plan – to set out long-term objectives and the means of achieving them.
8. Monitoring and assessment – to assess the condition of the forest, yields of forest products, chain of custody, management activities and their social and environmental impacts.
9. Maintenance of high-conservation-value forests – to maintain or enhance the attributes that define such forests.
10. Plantations – to reduce the pressures on and promote the restoration and conservation of natural forests.

4.12.2 Framework for RSM standards

Within the sustainability arena, globalisation of the supply chain is making it increasingly difficult to ensure that procurement takes into account all legal requirements together with environmental, social and economic considerations. There is therefore considerable demand for responsible sourcing standards operated by third parties to enable procurement to take into account major legislative and other requirements, as well as economic, environmental and social issues such as employment, safety, child labour and effects on local communities. Not surprisingly, there is now a proliferation of such standards on the world stage covering some or all of the major issues, which are mostly operated by first- or second-party certification schemes (see Section 3.6 for details of BRE Global's framework standard BES 6001 (BRE Global, 2008)).

* This list has been taken from the FSC website. Descriptions of principles are abbreviated.

4.13 WWF One Planet Future

www.wwf.org.uk/oneplanet

WWF has a campaign focus around the concept of a 'One Planet Future', where societal and individual demands on our planet are balanced with those of nature and the resources available, to ensure a sustainable and equitable future. The vision is based on a reduction in impacts from current levels, which equate to roughly three times the carrying capacity of the planet (based on the assumption that the world's current population will strive to live as we do in the UK), requiring the equivalent of three planets to maintain our current levels of consumption.

Changes are required throughout our global society and WWF has a number of areas of work ranging from individual ecological footprinting (WWF's Ecological Footprint calculator) to the concept of One Planet Homes. WWF lobbies governments and industry to implement system changes in the housing, transport, energy and food sectors.

Since 2003, WWF has been working with government, industry and consumers to move the concept of sustainable homes from the fringes to the mainstream of the UK housing sector. The campaign has been highly successful in raising awareness and has had a lasting and significant influence on the housing sector. WWF is now developing this work in the non-domestic sector.

4.14 GRI – Global Reporting Initiative

www.globalreporting.org

GRI's vision is that reporting on economic, environmental and social performance by all organisations should be as routine and comparable as financial reporting. GRI is a not-for-profit entity supported by members of its stakeholder networks.

The GRI Sustainability Reporting Guidelines form the basis of a Sustainability Reporting Framework that guides and supports organisations in identifying, measuring and disclosing their sustainability performance, and also provides stakeholders with a universally applicable, comparative framework, which provides the opportunity to benchmark performance and provide clarity in understanding disclosed information. The Reporting Framework provides transparency and accountability in all sizes of organisations and sectors across the world.

GRI is a worldwide, multi-stakeholder network with a robust governance structure to ensure consistency and transparency. Members from the business community and civil society, workers, investors, accountants and others all collaborate to develop and improve the tools and guidance through consensus-seeking approaches. The multi-stakeholder approach ensures the credibility and trust required of a global disclosure framework.

5 COMMON FEATURES OF ASSESSMENT TOOLS

5.1 Balanced scorecard

Basic rating systems for environmental/sustainability impacts score a series of sustainability issues, giving each a unit weight and then adding the individual scores to obtain an overall rating. While simple to use, the disadvantage of such an approach is that equal weight is given to each issue, eg for both an increment of energy performance and the provision of an ecological feature. In effect, a weighting is given to issues based entirely on the number of 'credits' attached to the issue, which makes it difficult to provide a balanced overall measure of performance while still encouraging best practice in each area.

Other systems use a 'balanced scorecard' approach to summing the relative performance of each issue and then add those 'weighted' scores to obtain an overall score. This approach aims to use the relative weightings to ensure that issues of key importance, such as reduction in energy use, are made more important in the overall score than other more easily and cheaply achievable objectives such as the provision of a cycle rack. That is not to say that both are not important in their own right, but if rating systems are to be readily accepted they have to pass a 'common sense' test in terms of the relative impact of each issue on the overall sustainability rating. In BREEAM, for instance, the relative weightings are chosen in consultation with industry and specialists so users of the system are likely to feel that the relative importance 'feels right', given our current state of knowledge of relative costs and impacts.

5.2 Choice of credits

Simple systems require fixed performance levels to be achieved for each sustainability issue so as to achieve an overall rating, and as such can be overly prescriptive. Such systems date rapidly and provide far less value for money in terms of sustainability achieved per unit of expenditure compared with their more sophisticated counterparts.

More advanced rating systems allow designers flexibility in achieving a target overall score by allowing them to trade, say, higher performance in materials selection against lower performance in water consumption. This approach has, on occasion, been criticised as it may, for example, theoretically allow the achievement of a high overall score without taking significant steps to reduce CO_2 emissions. When combined with a balanced scorecard approach to scoring, this theoretical risk does not occur in practice, and such an approach is essential if we are to achieve more sustainable outcomes without making those outcomes potentially unaffordable and in many cases unsustainable. This is because we do not know at any one time the relative costs of achieving each level of performance for each sustainability issue. It is clearly important that practitioners retain the flexibility of choosing which credits to achieve, and to what level, based on their creativity and the underlying costs, which vary over time and from market to market.

Having said that, it is clear that in some areas, and particularly with CO_2 emissions, there is a very strong consensus that we need to do much better than the regulatory minimum to achieve a high overall rating. Reflecting that, BREEAM 2008, for example, sets minimum mandatory energy performance levels corresponding to each grade of award, although it allows full flexibility between other issues.

For system designers, the introduction of such 'fixed' requirements is a very challenging step as it requires a thorough understanding of not only the technological possibilities to achieve each fixed requirement, but also the attendant cost. This is not a simple process in a rapidly changing market and requires continuous careful attention over time to keep fixed requirements achievable and relatively affordable in line with technological and economic developments. Experience in the UK with the CSH (CLG, 2008a, 2008b) shows that including a significant number of policy-related mandatory issues increases costs. For this reason, the number of fixed requirements in advanced systems should be limited.

5.3 Delivery in the marketplace

For the marketplace to provide advantage to those who subject their developments to the rating process, there clearly has to be certainty that public claims made regarding the sustainability rating of individual projects are justified. To achieve this, many systems require the award of a rating to be subject to 'certification' by an independent body.

The simplest and most common approach is for the body that owns the standard to offer to assess designs and award an appropriate rating certificate. This can be very expensive and bureaucratic, however, requiring the submission of large amounts of verifiable data, and is inherently inefficient as there is only one provider. This method has, in the past, encountered difficulty in adjusting capacity to meet market demand, although this can be mitigated through licensing multiple bodies to carry out the assessments. As this process has all of the assessment team in one place, it makes it significantly more of a burden to include a post-construction check. However, without a post-construction check there is considerable opportunity for 'specification downgrading' during construction.

A more flexible and accessible approach is to establish a network of individual licensed assessors who visit the design team and undertake assessments in the field, including a post-construction review to confirm that what has been built is what was specified. In addition to lower intrinsic costs and less bureaucracy, this method allows individual assessors and assessor organisations to compete on service and price in the marketplace, ensuring best-value service delivery for clients. As many more assessors can be trained than are necessary at any one time, this approach copes well with an expanding market. Assessors also provide 'real world' feedback from design teams and constructors to system designers to ensure systems remain relevant and practical.

Experience shows that the mode and economics of delivery can be just as important to successful system take-up as the standard itself.

6 CONCLUSIONS

There is a rapidly growing appetite for rating methodologies that can be used to demonstrate the environmental performance of our activities, ranging from personal carbon footprinting tools to complex sustainability assessments and standards of components, buildings and entire cities. There are also rapidly growing demands to demonstrate sustainability in many aspects of the built environment, which result in a flood of claims and counter-claims together with the development of more and more standards, guidance and rating methods. While much of this work is well founded and helpful in moving the agenda forward, the plethora of approaches introduces confusion and conflict in the marketplace and a lack of consistency in priorities and direction. This acts as a barrier to take-up and therefore to meeting the objectives that these initiatives set out to achieve.

It must be recognised that the construction industry is diverse and has many, often conflicting, commercial and policy objectives that can disrupt or divert the drive to greater overall sustainability. Organisations are increasingly using the sustainability label to promote their products in an ill-informed market with varying degrees of rigour and robustness. In his lucid, entertaining and insightful book, *Sustainable energy – without the hot air*, Professor David MacKay of the Department of Physics, Cambridge University, writes that the UK is suffering from 'an epidemic of twaddle about sustainable energy' and that we are 'inundated with a flood of innumerate codswallop' (2008). The same can be said about other environmental and sustainability issues.

Despite this, the UK is well ahead of any other country in terms of tackling this agenda, a fact that came out very strongly at the 2008 International Sustainable Building Conference SB08 in Melbourne (World Sustainable Building Conference, 2008). No other country has the spread of focused initiatives in place, which are increasingly well linked to policy. The drive to overcome the deficiencies that undoubtedly exist in these methodologies must not result in a loss of this hard-earned experience, knowledge or commercial position. There is a need to consolidate and, where possible, simplify the 'toolbox' to improve understanding and take-up, but experience and quality must not be lost. Neither must there be a 'dumbing down' of targets or a loss of the focused message for different stakeholders.

If we are to solve the problems of climate change we need, as Professor MacKay says, 'numbers – not adjectives' (2008), and we must base what we do on evidence, not public relations drivers. Many, but by no means all, of the initiatives in place follow this principle to some degree and this should be strengthened as we move forward. There is also a need to ensure consistency in the metrics used and the setting of common priorities and visions to ensure a common direction.

To this end, we hope that the UK construction industry will continue to take the lead in sustainable building and find a way of delivering cost-effective and comfortable zero carbon buildings well ahead of the government's targets. Our view is that this is best achieved by taking a strategic overview of the requirements for a sustainable built environment, feeding this back into the development of existing leading tools that are operating in the UK. Working to create and strengthen links, metrics and promotion of sustainability labels and guidance would have a dramatic effect both in terms of accessibility and in contributing to the development of government policy and industry strategies. The international dimension is paramount in taking this forward. Many client organisations, not to mention players in the construction and property sectors, do not restrict their operations to the UK or even European contexts. Increased international benchmarking and mapping of standards are vital. Drivers and needs vary considerably between climates, regulatory frameworks and, indeed, social and cultural priorities, and so there is no scope for a 'one size fits all' approach.

REFERENCES

Anderson J, Shiers D E and Sinclair M (2002). *The green guide to specification*. Oxford, WileyBlackwell, 3rd edition. (4th edition, BRE Report BR 501, due for publication in 2009.) Available from *www.brebookshop.com* or online from *www.thegreenguide.org.uk*

Atkinson C J and Butlin R (1993). *Eco-labelling of building materials and building products*. BRE Information Paper IP 11/93*. Bracknell, IHS BRE Press.

BRE (1990–2008)

Baldwin R, Leach S J, Doggart J and Attenborough M (1990). *BREEAM 1/90. An environmental assessment for new office designs*. BRE Report BR 183†. Bracknell, IHS BRE Press.

Crisp V et al (1991). *BREEAM 2/91. An environmental assessment for new superstores and supermarkets*. BRE Report BR 207†. Bracknell, IHS BRE Press.

Prior J J, Charlesworth J and Raw G J (1991). *BREEAM 3/91. New homes: an environmental assessment for new homes*. BRE Report BR 208†. Bracknell, IHS BRE Press.

Baldwin R et al (1993). *BREEAM 4/93. Existing offices: an environmental assessment for existing office buildings*. BRE Report BR 240†. Bracknell, IHS BRE Press.

Lindsey T (1993). *BREEAM 5/93. New industrial units: an environmental assessment for new office designs*. BRE Report BR 252*. Bracknell, IHS BRE Press.

Prior J J (ed) (1993). *BREEAM 1/93. New offices: an environmental assessment for new office designs*. BRE Report BR 234†. Bracknell, IHS BRE Press.

Prior J J and Bartlett P B (1995). *BREEAM New homes – Environmental standard: homes for a greener world*. BRE Report BR 278†. Bracknell, IHS BRE Press.

Yates A, Baldwin R, Howard N, Rao S (1998). *BREEAM 98 for offices: an environmental assessment method for office buildings*. BRE Report BR 350*. Bracknell, IHS BRE Press.

Howard N, Edwards S and Anderson J (1999). *BRE methodology for environmental profiles of construction materials, components and buildings*. BRE Report BR 370*. Bracknell, IHS BRE Press.

Prior J J (1999). *Sustainable retail premises: an environmental guide to design, refurbishment and management of retail premises*. BRE Report BR 366*. Bracknell, IHS BRE Press.

Anderson J and Howard N (2000). *The green guide to housing specification*. BRE Report BR 390*. Bracknell, IHS BRE Press.

Anderson J, Shiers D E and Sinclair M (2002). *The green guide to specification*. Oxford, WileyBlackwell, 3rd edition. (4th edition, BRE Report BR 501, due for publication in 2009.)

Rao S, Yates A, Brownhill D and Howard N (2003). *EcoHomes: the environmental rating for homes*. BRE Report BR 389*. Bracknell, IHS BRE Press.

Anderson J, Jansz A, Steele K, Thistlethwaite P, Bishop G and Black A (2004). *The green guide to composites*. BRE Report BR 475*. Bracknell, IHS BRE Press.

Bourke K, Ramdas V, Singh S, Green A, Crudgington A and Mootanah D (2005). *Achieving whole life value in infrastructure and buildings*. BRE Report BR 476*. Bracknell, IHS BRE Press.

BRE Trust and Cyril Sweett (2005). *Putting a price on sustainability*. BRE Trust Report FB 10*. Bracknell, IHS BRE Press.

Surgenor A and Butterss I (2008). *Putting a price on sustainable schools*. BRE Trust Report FB 15*. Bracknell, IHS BRE Press.

BRE (2005). *SAP 2005: the government's standard assessment procedure for energy rating of dwellings*. Published by BRE, Garston, and regularly updated. Latest version available from *www.bre.co.uk*

BRE (2006). *SBEM 2006: simplified building energy model*. Developed by BRE for CLG. Available from *www.bre.co.uk/energyrating*

BRE Global (2008). *BRE environmental and sustainability standard: framework standard for the responsible sourcing of construction products*. BES 6001: Issue 1.0. Garston, BRE Global, 2008.

British Standards Institution (2004). *Environmental management systems – General guidelines on principles, systems and supporting techniques*. British Standard BS EN ISO 14001:2004. London, BSI. Available from *www.bsi-global.com*

British Standards Institution (2008). *Quality management systems*. British Standard BS EN ISO 9001:2004. London, BSI. Available from *www.bsi-global.com*

Brundtland G H (ed) (1987). *Our common future: the World Commission on Environment and Development*. Oxford, Oxford University Press.

Communities and Local Government (2007). *Building a greener future: policy statement*. London, CLG, 23 July 2007. Available from *www.communities.gov.uk*

Communities and Local Government (2008a). *The Code for Sustainable Homes: setting the standard in sustainability for new homes*. London, CLG, February 2008. Available from *www.communities.gov.uk*

Communities and Local Government (2008b). *The Code for Sustainable Homes: technical guide*. London, CLG, October 2008. Updated regularly; latest version available from *www.communities.gov.uk*

Department for Business, Enterprise & Regulatory Reform (2008). *Strategy for sustainable construction*. HM Government in association with the Strategic Forum for Construction. June 2008. Available from *www.berr.gov.uk*

Directive 85/337/EEC (1985). *Assessment of the effects of certain public and private projects on the environment*. (Often referred to as the EIA or Environmental Impact Assessment Directive.) Official Journal of the European Communities L175, 5 July 1985; as amended by Council Directive 97/11/EC of 3 March 1997.

Directive 2001/42/EC (2001). *Assessment of the effects of certain plans and programmes on the environment*. Official Journal of the European Communities L197/30-37, 21 July 2001.

Directive 2002/91/EC (2002). *Energy performance of buildings*. Official Journal of the European Communities L001, 16 December 2002.

European Commission (2004). *Development of horizontal standardised methods for the assessment of the integrated environmental performance of buildings*. EC/DG Enterprise Standardisation Mandate M350 to CEN, 29 March 2004.

Henson R (2006). *The rough guide to climate change*. London, Penguin.

MacKay D J C (2008). *Sustainable energy – without the hot air*. Cambridge, University of Cambridge. Available from *www.withouthotair.com*

Milne R (2008). *Fresh zero carbon targets proposed in budget*. London, Planning Portal. Available from *www.planningportal.gov.uk*

Office of National Statistics (2008). *Construction statistics annual 2008*. London, ONS. Available from *www.statistics.gov.uk*

Pout C and MacKenzie F (2005). *Reducing carbon emissions from commercial and public sector buildings in the UK*. BRE Client Report No. 211 104. Produced under contract to DEFRA. Bracknell, IHS BRE Press.

Saunders T (2008). *A discussion document comparing international environmental assessment methods for buildings*. Available from BRE Global at *www.breeam.org*

Society of Environmental Toxicology and Chemistry and SETAC Foundation for Environmental Education (1991). *A technical framework for life cycle assessment*. Washington, SETAC.

Turner C (2008). *2016 – Resources for the journey to zero carbon*. Garston, National Centre for Excellence in Housing, p4. Available from *www.homein.org*

World Sustainable Building Conference (2008). Impacts of pollution in a changing urban environment. *Proceedings, World Sustainable Building Conference, Melbourne, Australia, 21–25 September 2008. www.sb08melbourne.com*

Notes

* BRE publications are available from *www.brebookshop.com*.

† Out of print, but available on special request from BRE Archive (breeam@bre.co.uk).

WEBSITES FOR TOOLS AND ORGANISATIONS

American National Standards Institute (ANSI)	*www.ansi.org*
Building Owners and Managers Association (BOMA)	*www.boma.org*
BRE Environmental Assessment Method (BREEAM)	*www.breeam.org*
Civil Engineering Environmental Quality Assessment and Award Scheme (CEEQUAL)	*www.ceequal.com*
Design Quality Indicator (DQI)	*www.dqi.org.uk*
DQI for Schools	*www.dqi.org.uk/schools*
ENVEST	*http://envestv2.bre.co.uk*
Forest Stewardship Council (FSC)	*www.fsc.org*
Global Reporting Initiative (GRI)	*www.globalreporting.org*
Green Building Initiative (GBI)	*www.thegbi.org*
Green Globes	*www.greenglobes.com*
Green Star	*www.gbca.org.au/green-star/*
Leadership in Energy and Environmental Design (LEED)	*www.usgbc.org/leed*
Sustainable Project Appraisal Routine (SPeAR®)	*www.arup.com/environment*
United Kingdom Accreditation Service (UKAS)	*www.ukas.com*
United Nations Environment Programme (UNEP)	*www.unep.org*
World Green Building Council (WGBC)	*www.worldgbc.org*
World Wildlife Fund One Planet Future	*www.wwf.org.uk/oneplanet*

GLOSSARY OF TERMS

ABE	Association of Building Engineers. *www.abe.org.uk*
BEE	Building Environmental Efficiency
BERR	Department for Business, Enterprise & Regulatory Reform. *www.berr.gov.uk*
BIP	Building Information Pack. A collection of documents that includes details of a building and its systems to inform future building owners, occupiers and maintainers.
BRE	Building Research Establishment. *www.bre.co.uk*
BRE Trust	Owner of the companies in the BRE Group. The BRE Trust is a registered charity aiming to advance knowledge, innovation and communication in all matters concerning the built environment. *www.bretrust.org.uk*
BREEAM	BRE Environmental Assessment Method. The BREEAM family of assessment methods and tools is designed to help construction professionals understand and mitigate the environmental impacts of all types of developments. BREEAM Buildings can be used to assess the environmental performance of any type of building (new or existing). See Section 3.1. *www.breeam.org*
BREEAM International	International version of BREEAM that can be tailored to local conditions including climate, regulations and markets. See Section 4.4. *www.breeam.org*
BSI	British Standards Institution. *www.bsi-global.com*
CABE	Commission for Architecture and the Built Environment. *www.cabe.org.uk*
CASBEE	Comprehensive Assessment System for Building Environmental Efficiency. Japanese methodology for calculating a Building Environmental Efficiency score. Constitutes a self-assessment check system for raising the environmental performance of buildings. Five different ratings are available. See Section 4.7. *www.ibec.or.jp/CASBEE/english*
CEEQUAL	Civil Engineering Environmental Quality Assessment and Award Scheme. An assessment method and reward scheme for civil engineering schemes. CEEQUAL complements BREEAM by providing a means of evaluating the environmental quality of the procurement process beyond buildings and communities, but focuses on robust processes and target-setting procedures within a project rather than setting absolute targets as BREEAM does. See Section 3.4. *www.ceequal.com*
CEN	Comité Européen de Normalisation (European Committee for Standardisation). *www.cen.eu*
CIC	Construction Industry Council. *www.cic.org.uk*
CIRIA	Construction Industry Research and Information Association. *www.ciria.org.uk*
CLG	(Department for) Communities and Local Government (replaced ODPM in 2006). *www.communities.gov.uk*
Co-product	A product that results from the manufacturing process, use or disposal of another product. The co-product has monetary value and can be sold.
CPA	Construction Products Association. *www.constructionproducts.org.uk*
CSA	Canadian Standards Association. *www.csa.ca*
CSH	Code for Sustainable Homes, launched in December 2006. Sets mandatory minimum standards against energy, water, construction and household waste, materials and lifetime homes that relate to key government targets and policies. It has six potential star ratings. See Section 3.2. *www.communities.gov.uk*
DEC	Display Energy Certificate. This shows the actual energy use of a building based on the energy consumption recorded by gas, electricity and other meters. It is required from 30 December 2008 in buildings of floor area over 1,000 m^2 that are occupied by a public authority, an institution providing a public service or one frequently visited by members of the public.
DEFRA	Department for Environment, Food and Rural Affairs. *www.defra.gov.uk*
DQI	Design Quality Indicator. A tool to measure the design quality of buildings. See Section 3.3. *www.dqi.org.uk*
EC	European Commission. *http://ec.europa.eu*

EcoHomes BRE Environmental Assessment Method applied to housing. *www.breeam.org*

EMAS The European Community Eco-Management and Audit Scheme is a voluntary environmental management system for organisations operating in the European Union and the European Economic Area. Its purpose is to enable companies and organisations of both private and public sectors to manage their environmental impacts in a systematic way.

ENVEST A software tool that simplifies the complex process of designing buildings with both low environmental impact and whole life costs. The current version, ENVEST 2, allows both environmental and financial issues to be optimised by a client to achieve best value. See Section 3.5.3. *http://envestv2.bre.co.uk*

Environmental Profile A Life Cycle Assessment-based evaluation of the environmental impacts arising from the extraction/manufacture of construction materials carried out under the BRE Environmental Profiling methodology. See Section 3.5.1.

EPC Energy Performance Certificate. This is required for all buildings at the point of construction, sale or rental. It records the energy efficiency and carbon emissions of a building. It shows a rating from A to G, where A is very efficient and G is very inefficient.

FSC Forest Stewardship Council. An independent, non government, not-for-profit organisation established in 1990 to promote the responsible management of the world's forests. It provides standard setting, trademark assurance and accreditation services for companies and organisations interested in responsible forestry. See Section 4.12.1. *www.fsc.org*

GBCA Green Building Council of Australia. *www.gbca.org.au*

GBI Green Building Initiative. A not-for-profit organisation whose mission is to accelerate the adoption of building practices that result in energy-efficient, healthier and environmentally sustainable buildings by promoting credible and practical green building approaches for residential and commercial construction. *www.thegbi.org*

GBTool Green Building Tool (GBTool; now known as the SBTool) was designed to assess the environmental and sustainability performance of buildings in 1996 to provide a common framework for the presentation and evaluation of buildings case-studied as a part of the Green Building Challenge (GBC). It has been updated on a regular basis for biannual GBCs. The GBC process was launched by Natural Resources Canada, but responsibility was handed over to the International Initiative for a Sustainable Built Environment (iiSBE) in 2002.

GDP Gross domestic product

Green Globes Environmental assessment and certification scheme based on BREEAM, incorporating self assessment with independent third-party verification. See Section 4.3. *www.greenglobes.com*

Green Guide Provides environmental ratings of construction elements based on Life Cycle Assessments. It is available in printed form (see References) and online at *www.thegreenguide.org.uk*. See Section 3.5.2.

Green Star An environmental assessment method for buildings derived from BREEAM. Green Star is similar to BREEAM, but reflects important differences between Australia and the UK such as climate, local environments and construction industry standard practices. See Section 4.6. *www.gbca.org.au/green-star/*

GreenPrint A methodology developed to help design teams to deliver masterplans that maximise the potential for sustainable communities. It can be applied to a wide range of development types from urban extensions to business parks. The GreenPrint methodology provides a unique assessment for each project, in order to maximise its sustainability potential, and is designed in consultation with the client and key stakeholders. *www.bre.co.uk*

GRI Global Reporting Initiative. Works towards making reporting on economic, environmental and social performance by all organisations as routine and comparable as financial reporting. See Section 4.14. *www.globalreporting.org*

HIP Home Information Pack. A collection of documents that any vendor selling their house must provide to the prospective buyer. *www.explorehomeinformationpacks.co.uk*

iiSBE International Initiative for a Sustainable Built Environment. iiSBE is an international, not-for-profit organisation whose overall aim is to actively facilitate and promote the adoption of policies, methods and tools to accelerate the movement towards a global sustainable built environment. *www.iisbe.org*

LCA Life Cycle Assessment. A compilation and evaluation of the inputs, outputs and the potential environmental impacts of a product system through its life cycle (*Environmental management – Principles and framework*. International Standard ISO 14040:2006).

LEED Leadership in Energy and Environmental Design. It was established by the US Green Building Council to improve the way the construction industry assesses sustainability issues by providing a simple, easy-to-use label. Four ratings are available depending on performance: Certified, Silver, Gold and Platinum. See Section 4.5. *www.usgbc.org/leed*

ODPM Office of Deputy Prime Minister (replaced by CLG in 2006)

OGC Office of Government Commerce, part of the UK Treasury. *www.ogc.gov.uk*

PEFC Programme for the Endorsement of Forest Certification. The PEFC Council is an independent, not-for-profit, non-government organisation, founded in 1999, which promotes sustainably managed forests through independent third-party certification. The PEFC provides an assurance mechanism to purchasers of wood and paper products that they are promoting the sustainable management of forests. *www.pefc.co.uk*

PPG Planning Policy Guidance Notes are prepared by the government to provide guidance to local authorities and others on planning policy and the operation of the planning system. Local authorities must take their contents into account in planning decisions. These have been replaced by Planning Policy Statements.

PPS Planning Policy Statements replace the earlier Planning Policy Guidance Notes and serve the same purpose.

Private wire arrangement Defined in the Code for Sustainable Homes as '... [an arrangement] where any electricity generated on or in the vicinity of the site is fed directly to the dwellings being assessed, by dedicated power supplies'.

RSM Responsible Sourcing of construction Materials. BRE, in conjunction with CPA, BSI and others, has developed and published a framework standard for the Responsible Sourcing of Construction Products, BES 6001. Key players in the construction materials and components sector are currently working with BRE to develop BES 6001-compliant, sector-specific standards. See Sections 3.6 and 4.12.

SBA Sustainable Buildings Alliance. Established to provide sustainable solutions to companies committed to achieving sustainable real estate objectives. See Section 4.10. *www.sustainablebuildingsalliance.org*

SETAC Society of Environmental Toxicology and Chemistry. See Section 4.11. *www.setac.org*

SFI Sustainable Forestry Initiative®. This is an independent, charitable organisation dedicated to promoting sustainable forest management in Canada through the operation of forestry certification schemes. *www.sfiprogram.org*

SMARTWaste SMARTWaste is a suite of tools and consultancy services developed by BRE. It can be applied to any waste-generating activity in the construction and demolition industries. *www.smartwaste.co.uk*

SPeAR® Sustainable Project Appraisal Routine. Developed by Arup to demonstrate the sustainability of a project, process or product to be used either as a management information tool or as part of a design process. It works on a four-quadrant model based on environmental protection, social equity, economic viability and efficient use of natural resources. *www.arup.com/environment*

UKGBC UK Green Building Council. *www.ukgbc.org*

UNEP United Nations Environment Programme. UNEP's mandate is to co-ordinate the development of environmental policy consensus among member states by keeping the global environment under review and bringing emerging issues to the attention of governments and the international community. It reports to the UN General Assembly through the Economic and Social Council. See Section 4.9. *www.unep.org*

USGBC US Green Building Council. *www.usgbc.org*

World GBC The World Green Building Council movement is a union of national councils with the mission to accelerate the transformation of the global built environment towards sustainability. World GBCs represent over 50% of global construction activity associated with more than 15,000 companies and organisations worldwide. See Section 4.8. *www.worldgbc.org*

WWF World Wildlife Fund. Aims to conserve biodiversity and address threats to the environment by working with people for sustainable solutions. *www.wwf.org.uk*

WWF One Planet Future A campaign by WWF to bring people together to make changes to their lives by inspiring individuals, businesses and government to contribute jointly to the reduction of environmental impacts by moving from a three planet lifestyle to a one planet lifestyle. See Section 4.13. *www.wwf.org.uk/oneplanet*